TABLE OF CONTENTS

FOREWORD

In writing this book, Helen McGonagle has made great use of the treasure trove found, as she explains, in a strong room in Cork City Library: a set of annual Reports of the city's Carnegie Free Library. This source material is all the more valuable because, like a large section of the city centre, that library was lost on the December night, in 1920, when Cork was burned.

Helen hones in on evidence relating to the 'Ladies Reading Room', and there is much in her work to remind us of both the interests of women and their status in this period. We find, for example, that the 'Juvenile Library' was initially known as the 'Boys' Room' and that few women entered the public library before that separate room was opened for them, with those who did flitting through with an air of embarrassment.

The book provides insights into one aspect of women's experience in a period when the women's movement became increasingly active in its demand for the vote. In October 1910, a branch of the Irish Women's Franchise League was set up in Cork but within months the members established their own, non-militant as well as non-party, Munster Women's Franchise League. With the focus on the vote, women with differing political outlooks, like Máire MacSwiney and writer Edith Somerville, worked together to argue that women must have the vote because their perspectives, particularly on social issues, were essential to the political system. In these days of mass media we forget not only the importance of libraries as sources of information and education but also that women like these had to take their message out to into streets and halls and defend it against anti-suffragists.

During the decade covered, the Irish Language, Nationalist, Labour and Women's Movements heralded an era of new ideas, high hopes and political excitement. With her focus on the Ladies's Reading Room, Helen adds a fresh layer to our growing understanding of Cork City's social and cultural history.

Dr. Sandra McAvoy,
Department of Women's Studies, University College Cork.

INTRODUCTION

A Room of their Own:Cork Carnegie Free Library & its Ladies Reading Room 1905 -1915 is the third book to appear in the last decade on the history of public libraries in the city of Cork. The first – *A Grand Parade* – was published in 2005 on the 150th anniversary of the Public Libraries (Ireland) Act, and took a broad brush approach in recounting the establishment and development of the library service over more than one hundred years. The second – Thomas McCarthy's *Rising from the Ashes* – told the heartening story of how the city's civic leaders rebuilt the library and its collections, after the catastrophic destruction of the Cork Carnegie Free Library by Crown Forces in December 1920.

This new book is focused on one important aspect of the Cork Carnegie Free Library. Having sketched a picture of the Edwardian city of Cork in the early 20th century, *A Room of their Own* recounts the development of the Carnegie building, which opened to the public in 1905. Helen McGonagle shines a light on the views which were prevalent at the time, in the United Kingdom and in North America, about the benefits of separate reading rooms for women. Her insights on library annual reports – why they were produced, how they were used then (or not), and what uses they might serve today – are particularly interesting.

The core of Helen McGonagle's book is her discussion of the Ladies' Reading Room at the Cork Carnegie Free Library – how it provided for its women readers in the first decades of the 20th century, and how women used the Library and its resources.

Cork City Libraries are delighted to publish Helen McGonagle's *A Room of their Own*. It is a unique contribution to library history, Irish and international, and a very valuable contribution to the social and cultural history of a city which would play such a pivotal role in Ireland's struggle for independence, just a few years after the period which Helen McGonagle brings to life.

Liam Ronayne
Cork City Librarian

CHAPTER ONE
A BUSTLING EDWARDIAN CITY
AN INTRODUCTION

The idea for this work came from the discovery of the small collection of Carnegie Free Library Annual Reports dating from 1905-1915 in the strong room of the City Library in Cork. It is a fascinating collection, containing a myriad of information about life in Cork at that time. Little or no use has been made to date of these reports as a research source, perhaps because the focus of library history in Cork has tended to be on the burning of the Carnegie Library, along with much of the city centre by Crown forces in 1920, and the subsequent effort to restock and rebuild the Library. While public libraries have evolved into one of the most popular public services in modern Britain and Ireland, historians in general have tended to overlook the wider historical significance and context of these institutions and their annual reports.[1] This may be explained to some extent by the relative inaccessibility of these reports, held as they are, in the main, by individual library authorities.

How libraries developed has been explored to some extent, for example Kelly's *History of Public Libraries in Great Britain 1845-1975,* which is a chronological record of fact, but little consideration has been given by library historians as to why libraries developed as they did. This work attempts in a small way to do this, by adding to the body of library historiography in Ireland and in Cork in particular and placing annual reports and public library developments in general within what Black (1996) terms the 'field of cultural history' [2]

The Carnegie Free Library in Cork with Librarian Mr. Wilkinson and his wife on the balcony. Image courtesy of the Lawrence Collection (1.1)

Historic annual reports in general, although varying in length, follow a standard template providing:

- a narrative commentary on a particular library's progress during the period of the report,
- statistical information about these institutions with regard to stock held, numbers and type of books issued
- number and profession/trade (if any) of borrowers.

These reports were usually signed either by the chairman of the library committee or the librarian on behalf of the Library Committee[3].

Main Hall, Great Cork Exhibition 1902. Image courtesy of the Lawrence Collection (1.2)

Early Twentieth Century Cork

To appreciate the value of the information in the Library Annual Reports, it is necessary to imagine Cork in the early 1900's. It was at least on the surface, a city at peace with itself, a busy Edwardian city with a bustling port and a well established merchant class[4]. The Cork International Exhibition, visited by Edward VII and Queen Alexandra, had taken place on the Mardyke, along the banks of the river Lee during 1902 and 1903 and was attended by tens of thousands of visitors both national and international.

Wealthier merchants and members of the prosperous predominantly Catholic middle classes had by now moved from the narrow streets and laneways of the inner city to the outer reaches of the city at Sunday's Well, Montenotte, Tivoli, Blackrock and other suburbs[5]. However, this affluence and success must be contrasted with the appalling conditions endured by those not so fortunate. The houses vacated by the middle classes became tenement homes for the working classes and the unemployed. Living conditions in many of these tenements were generally 'poor, overcrowded and insanitary'[6] with tenement buildings accommodating up to 23 families[7].

Housing in Cork. Image courtesy of the Lawrence Collection (1.3)

Attempts by Cork Corporation and the Improved Dwellings Company to clear these slums and to provide alternative housing for the occupants were only partially successful. While some slums were demolished, not enough houses were built to accommodate the former slum dwellers. The rents charged for the new houses were often out of the reach of the poorest families[8]. In addition, both Cork Corporation and the Improved Dwellings Company tended to favour tenants with what they considered to be prudent and temperate habits. This was reflected in the names chosen for some of the new housing schemes - Industry Street and Prosperity Square, situated near Barrack Street, are two good examples[9]. This resulted in more desperately poor people being crowded into the remaining tenements.

Political energy in Cork, as in the rest of Ireland at the time was largely concentrated on the nationalist struggle to achieve Home Rule rather than on achieving improved working conditions and rights for workers. On 28 September 1914 Asquith's Home Rule Bill became law with the support of the Irish Parliamentary Party led by John Redmond. Its provisions were immediately delayed until 'not later than the end of the present war'[10] – the war was expected to be over within a matter of months. Redmond, William O' Brien and other nationalist leaders called for Irish support for the war while the more radical wing of the nationalist movement voiced its opposition to any Irish men fighting for British forces[11]. This difference in attitude towards the war led to a split in the nationalist Irish Volunteer movement in Cork as it did in the rest of the country. In general however, support for the war was widespread in Cork. Many men volunteered for the army and organisations were set up to support the troops, the wounded and the families of those in the armed forces. For a time, divisions between nationalists and unionists appeared to be forgotten.

Since the introduction of the Education Act in 1831, primary education was available for all regardless of class or creed. Originally, non-denominational education was planned, but it quickly became segregated and came under the control of the religious congregations, both Catholic and Protestant[12]. The curriculum focused on teaching basic reading, writing and arithmetic to all. However for girls, the focus quickly changed to domestic tasks, preparing them for becoming wives and mothers[13].

The Gaelic Revival

The years 1905 – 1915 encompassed a time of great change in Ireland. New movements stressed the importance of Irish identity, Irish race and Irish culture. While the majority of people in Ireland still supported the Irish Parliamentary Party in its quest for Home Rule for Ireland, the late 19th and early 20th century saw a considerable Gaelic revival and the suggestion that perhaps Home Rule was not enough. This revival of cultural nationalism was particularly strong in the cities of Cork and Dublin and it was the leaders of this movement who provided much of the leadership of the Republican revolution which followed[14]. The Gaelic revival spawned a renewed emphasis on all things Gaelic and nationalist – from fairy tales, drama and poetry to music, the Gaelic language and sports.

Much of this revival was promoted through writing – in the form of novels, plays, poetry, newspaper and journal articles, published letters and speeches which were later reported in newspapers and journals. This renaissance of literature and drama was lead by poets such as W.B. Yeats, Eva Gore Booth and A.E (George Russell) and dramatists such as Lady Gregory and J.M. Synge. Indeed Senia Paseta in her 2013 work *Irish Nationalist Women* claims that political literature was probably the single most influential politicising agent of nationalist women of that era, with many women discovering politics through books, newspapers and journals.[15] The fledgling free library service was instrumental in providing this reading material to the people of Ireland.

Lord Mayor's Room.
Municipal Buildings.
Cork.

October 26th, 1901.

June 3/02

4

Dear Sir,

On behalf of the citizens of Cork I beg to apply to you for a donation with a view to establishing a free public library for the great and important City of Cork. I am encouraged to do so by the unbounded generosity you have shown to other places, and believing that what you demand of us, to secure its permanent retention for the citizens, could be satisfactorily provided. I hope, amongst the many other monuments which you are erecting throughout the United Kingdom, that we here in Cork could be, also, able to point to one as a mark of your generosity to the City. While there was some slight provision made a few years ago to provide one or two rooms for a free library they are utterly inadequate for the wants of our important City, and if we were favoured with your donation the few rooms, available at present, would be very much needed for technical instruction purposes.

Yours faithfully,

Edward Fitzgerald

Lord Mayor of Cork.

The large increase in the number of the visitors to the Library, **8,794**, or from 242,940 to **251,734**, has necessitated some changes both in the Newsroom and Reference Reading Room, so as to make the most of the available space. As a result of the re-arrangement it has been possible to set aside two tables for the exclusive use of ladies. The placing of pads on the legs of the chairs has increased the comfort of readers by diminishing the noise.

The result of these alterations has raised the cost of fittings and repairs to £50 17s. 9d., an increase of £38 17s. 4d., and the year closed with a debtor balance of £12 3s. 8d.; however, the Committee consider the additional outlay has been fully warranted by the increased comforts provided for those using the Library.

Copy of application letter from Edward Fitzgerald, Lord Mayor of Cork with supporting extract from Annual Report. Fitzgerald Park in Cork is named after him. (1.4)

CHAPTER TWO

'A FREE PUBLIC LIBRARY

FOR THE GREAT AND IMPORTANT
CITY OF CORK'[16]

Cork has long had significant collections of books for both learning and pleasure, from the seventh-century monastic school associated with St. Finbarr at Gillabbey near the present University site through to the emergence of various institutions in the early to middle nineteenth century, such as the Cork Literary and Scientific Society, the Royal Cork Institute, and the Cork Cuvierian Society. It was not until the last decade of the nineteenth century, however, that the public library service began in the city. While Cork was the first of the Irish cities to adopt the Public Libraries (Ireland) Act of 1855, which allowed municipal councils to establish a library and charge an annual rate of a maximum of one penny on city businesses to support it, difficult financial circumstances meant that a library service was not opened until December 1892, in what is now the Crawford Municipal Art Gallery[17].

Application for a gift from Andrew Carnegie
Application was made in October 1901 to the well-known benefactor of libraries Scottish-born Andrew Carnegie by the Lord Mayor of Cork, Edward Fitzgerald, for a donation 'with a view to establishing a free public library for the great and important City of Cork'.

£10,000 was gifted to the people of Cork by Mr. Carnegie in August 1902 for the purpose of building a new free library in the city.

7

If maximum rate were devoted to Library,
revenue would be £680

say ~ a £10000 building

FREE PUBLIC LIBRARY.

Town? **Cork**

Population? (last census) **75,978**

Has it a Library at present? **Yes**

How Housed? **In a portion of the Schools of Science and Art Nelson's Place**

Expenditure for Support Last Year? **£610.14.2**
See Page 7 of report sent herewith

Has Library Act been Adopted? **Yes**

If so, what is the Revenue under it? **£523.3.5 per annum**

If not, will it be Adopted? _____

Is Requisite Site Available? **Yes**

Amount now Collected toward Building? _____

To facilitate Mr. Carnegie's consideration of your appeal, will you oblige by filling in the above?

Respectfully,

Jas. Bertram,
Secretary.

Lord Mayor's Room,
Municipal Buildings,
Cork.

17th September, 1903.

Andrew Carnegie, Esq. LLD
Skibo Castle,
Dornoch. Sutherland.

Dear Sir,

The date agreed upon for conferring upon you the Honorary Freedom of our City, and laying the foundation stone of the Free Library which you are generously enabling us to build, being now little more than a month distant, we are completing our arrangements for the auspicious double event.

May I therefore ask you and Mrs Carnegie to do me and the Lady Mayoress the honor of lunching with us on the occasion, so that we may invite the members of the Municipal Council and other leading citizens to meet you? I have no doubt that my fellow-citizens will esteem it a high privilege to do so.

Believe me, dear Sir,

Very truly yours

Edward Fitzgerald
Lord Mayor

7th April, 1903.

D. F. Giltman, Esq.,
Municipal Buildings, Cork, Ireland.

Dear Sir,

Mr. Carnegie is deeply appreciative of the Lord Mayor's desire and would be glad to have Mrs. Carnegie lay the stone if it could be done next October. Mr. Carnegie's plan is to leave Skibo in October and go direct to Ireland, visiting Waterford and Limerick as he has promised to do, and then proceeding to Cork, afterwards sailing for United States from Queenstown.

Respectfully yours,

P. Secretary.

Clockwise from left
Application for funding
with handwritten notes (2.1)

Invitation to lunch from
Lord Mayor to Mr & Mrs Carnegie (2.2)

Letter setting out the
Carnegies' travel plans (2.3)

22nd May, 1903.

D. F. Giltnan, Esq.
Municipal Buildings, Cork.

Dear Sir,

Mr. Carnegie has received your letter of 27th
April and asks me to advise you that he proposes to
sail on the Cedric, joining at Queenstown on 24th
October, and that therefore the 23rd would be the date
on which he could attend to lay the foundation stone
and receive the proposed honor at the hands of Cork.

Respectfully yours,

P. Secretary.

Lord Mayor's Room,
Municipal Buildings,
Cork.

27th May, 1903

Dear Sir,

I am directed by the Lord Mayor to
acknowledge your letter of 22nd instant,
intimating that Mr. Carnegie would visit
Cork on the 23rd October, to lay the Foundation
Stone of the new Free Library and receive
the Honorary Freedom of the City.

Would you kindly draw Mr. Carnegie's
attention to the fact that the 23rd will be a
Friday, and ask if it would suit his convenience
to substitute Thursday the 22nd, or any other
day of the week for same.

Yours faithfully,

D. Giltinan

James Bertram, Esq.
Private Secretary
Skibo Castle,
Dornoch,
Sutherland.

Clockwise from left
'Suggested travel dates from
Mr. Carnegie's secretary (2.4)

Letter pointing out that the 23rd is
a Friday from The Lord Mayor (2.5)

Confirmation of 22nd October (2.6)

Confirmation Telegraph (2.7)

A. POST OFFICE TELEGRAPHS.
(Inland Telegrams.)   For Postage Stamps.

NOTICE.—This Telegram will be accepted for transmission subject to the Telegraph Acts, the Regulations made thereunder, and the Notice printed at the back hereof.

TO { Giltinan Municipal Buildings Cork

Mr Carnegie finally decided come
Cork Twentysecond October as
suggested

FROM { Bertram

The Name and Address of the Sender, IF NOT TO BE TELEGRAPHED, should be written in the Space provided at the Back of the Form.

46

Lord Mayor's Room,
Municipal Buildings,
Cork.

19th June, 1903

Jas Bertram, Esq.
Private Secretary to Dr Carnegie.
Skibo Castle,
Dornoch, Sutherland.

Dear Sir,

I duly received your wire
on 15th instant, and am directed by the
Lord Mayor to ask you to convey to
Mr Carnegie his thanks for having finally
decided on Thursday 22nd October next
as the date of his visit to Cork.

Yours faithfully,

D. Giltinan
Sec.

As a way of thanking him for his generosity the Lord Mayor invited Mr. Carnegie and his wife to Cork to lay the foundation stone of the library and to receive the Freedom of the City. It appears that some difficulty arose with coming to an agreed date, the first date suggested by Mr. Carnegie was 23 October, 1901 and was rejected by the Lord Mayor because it was a Friday, (probably because this was a fast day), a date in July was then suggested and rejected again by the Lord Mayor because plans for the October visit were already in train. October 21 1903 was finally agreed on.

Detailed correspondence between Mr. Carnegie's secretary James Bertram and the Lord Mayor's secretary D. F. Giltinan record his travel arrangements and accommodation requirements. His timetable in Ireland saw him visit Dublin, Waterford, Limerick and Cork before departing for New York from Queenstown (Cobh) on board the 'Cedric'.

September 29th, 1903.

D. F. Giltinan, Esq., Sec.,
   Lord Mayor's Room, Cork.

Dear Sir,

We are much obliged for your letter of 25th September and would be still further obliged if you will kindly secure accommodation at the most preferable hotel among those you name for the time during which Mr. Carnegie is to be in Cork. - A sitting-room and bed-room for Mr. Carnegie, a bedroom for myself, and a bedroom for Mr. Carnegie's man-servant who will accompany him.

Mr. Carnegie notes the train you mention, arriving at Cork at 2 P.M. Of course you know best, but he hopes this will give time for your functions.

Respectfully yours,

P. Secretary.

P.S. - Mr. Carnegie does not bear arms.

The Imperial Hotel on Pembroke Street was recommended as being of 'the highest class'[19] by Mr. Giltinan and rooms were reserved there – A sitting room and bed room for Mr. Carnegie, a bedroom for Mr. Bertram, his secretary and a bedroom for his man-servant.

The foundation stone of the new library was laid by Andrew Carnegie and he was made a freeman of the city in ceremonies which took place on 21 October 1903. He spent his last night in Cork at the residence of the Bishop of Cloyne, Robert Browne (uncle of the photographer Fr. Francis Browne S.J.) in Queenstown.

Mr. Carnegie's speech at Cork references Home Rule and expounds on his ideas of where Britain and Ireland should give their allegiances. He suggests that Britain and Ireland should join with the United States of America – 'there

refreshment before attending the Council, which will be called for 12 o'clock. The proceedings in Council, presenting the "Freedom", would not occupy an hour, and would be immediately followed by the laying of the foundation stone of the Library, the site of which immediately adjoins the Municipal Buildings.

This would not take half an hour, and the Luncheon would take place in the Municipal Buildings at 1.30, and all would be over soon after 3 o'clock.

If after that, Dr Carnegie wished to see the City, and more particularly the Exhibition, the Lord Mayor would be most happy to place himself at his disposal —

58

for the purpose.

Thus you will see, the programme need not at all entail any late hours upon Dr. Carnegie, whom we should be very far from wishing to subject to any undue fatigue.

The Lord Mayor's anxious desire would be to make his visit to Cork as easy and pleasant as possible.

In the matter of hotel accommodation, if I were to choose that of the highest class I should engage rooms for you at the "Imperial", but if something quieter were preferred, I should recommend Moore's Family Hotel, on Morrison's Quay, which is apart from the noise and bustle of the mid-city.

Kindly let me know your wish on this point as soon as convenient.

I trust you for the reference books about Cork which I sent you on the 1st instant.

Yours faithfully

D.J. Giltinan
Lord Mayor Secretary.

15 Oct

Correspondence re Mr Carnegie's accommodation in Cork (2.8, 2.9)

is no reason why Ireland, Scotland and England should not be in an English speaking Union with the other branch of the race'[20]

Request for more money

In September 1904, Augustine Roche who was then Lord Mayor of Cork wrote to Andrew Carnegie, setting out a statement of the building works and costs incurred so far

He remarks that a sum of £1,000 in addition to that already provided will be needed for furnishing, heating, etc. On 7 October 1904 Mr. Carnegie's secretary replies saying that the extra £1,000 will be provided.

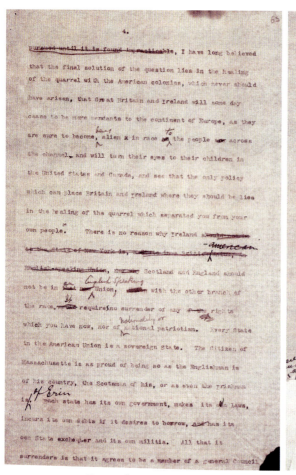
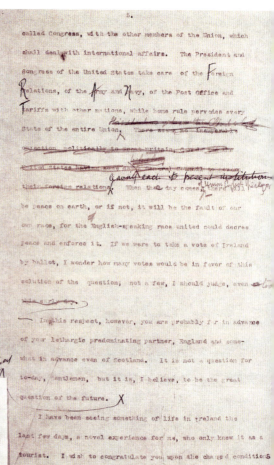

Extract from Mr. Carnegie's speech (2.11)

Lord Mayor's Room,
Municipal Buildings,
Cork.

9th September, 1904.

Andrew Carnegie Esq, L.L.D.,
    Skibo Castle,
        Dornoch, Sutherland.

Dear Sir,

The erection of the Free Library Building for this City, towards the cost of which you so generously made a grant of £10,000, has now reached a stage at which the Corporation think it right to lay before you a short statement.

As stipulated, the Corporation made a free grant of a valu - able site (adjacent to the Municipal Buildings), upon which some portion of the Corporation offices previously stood, and they have cleared the site at a cost of about £150.

The preparation of plans, quantities, advertising, etc, pre - liminary to making a contract cost £159.12.4.

A Contract was then entered into, the amount of which with necessary extras, to date, is - £9,490.3.3.

To this are to be added the salary of a Clerk of Works for the whole term of the contract, - £170.0.0.; and sundry incidental expenses £17.5.7.

Those items (exclusive of providing and clearing the site) tot up to £9,817.1.2., the expenditure of which will give the City the new

(1)   Building

Building without any of the necessary accessories.

Provision has now to be made for the Heating, estimated at about £300, furnishing, say - £400; and Miscellaneous works and fittings, estimated at say £200.

The Cork Gas Consumers' Company has undertaken to provide and fix the necessary pipes and fittings for lighting the building and the approach thereto, free of charge.

Excluding the site, the expenditure properly coming under the head of erecting and fitting up the building stands thus

Plans, quantities and advertising	£159.12.4.
Contract and extras to date	9490. 3.3.
Clerk of Works	170. 0.0.
Incidental expenses	17. 5.7.
Heating and ventilating (say)	300. 0.0.
Furnishing (say)	400. 0.0.
Miscellaneous works and fittings (say)	200. 0.0.
Total cost of buildings	10,717. 1. 2.

or £717.1.2. (which may probably reach £1000) in excess of your hand - some gift of £10,000.

The only source of income which the Corporation can apply to the purposes of a Public Library is the Penny Rate, which produces about £700, but this is already charged with the balance of a loan of £2,000, in connection with the old library building, which is being

(2)   repaid

---

repaid at the rate of £200 every three years, with 4 per cent interest; and you will no doubt remember that you made it a condition of your grant, that the proceeds of the penny rate should be devoted to the up-keep of the library, for which it is barely sufficient.

Under these circumstances the Corporation have asked me to acquaint you respectfully with the fact that a sum of nearly £1,000, in addition to your grant, has to be provided before the new edifice, with which your honored name is associated, can be ready to have the library transferred to it from its present unsuitable habitation; and that they have at present no fund at their disposal from which the deficien -cy can be made up.

I have the honor to be,
    My Dear Sir,
        Yours Very Truly,
            Augustine Roche
                Lord Mayor.

(3)

Statement regarding the progress of building work and a request for an extra £1,000 from the Lord Mayor (2.12)

Note agreeing to provide extra £1,000 from Mr. Carnegie's secretary (2.13)

October 7th, 1904.

Augustine Roche, Esq.,
    Lord Mayor of Cork.

Dear Sir,

Mr. Carnegie will be glad to provide the Thousand Pounds required to complete the Library Building and has authorized his Cashier to credit you with the increased amount.

Respectfully yours,

P. Secretary.

The Lord Mayor,

(ALDERMAN JOSEPH BARRETT,)

Requests the pleasure of the Company of

Doctor Andrew Carnegie

*at the Opening of the Carnegie Free Library, on
Tuesday, the 12th September, 1905, at 2.30 p.m.,
and at Luncheon in the Municipal Buildings
immediately after.   An early answer will oblige.*

Please present this Card at Luncheon Room.

Municipal Buildings, Cork.

September 4th, 1905.

Joseph Barrett, Esq.,
    Lord Mayor of Cork.

Dear Sir,

    Mr. Carnegie desires me to thank you for the kind
invitation conveyed in yours of the 1st inst. to be present
at the opening of the new Library Building and regrets that
his engagements here will make it impossible for him to be
present.

        Respectfully yours,

            P. Secretary.

Top:
Invitation to the Official Opening
of the Carnegie Free Library,
Tuesday 12 September 1905
(2.14)

Left:
Mr Carnegie was unable
to attend (2.15)

Official Opening

The opening ceremony, took place on September 12th, 1905, at 2.30 pm, conducted by the Right Hon. The Lord Mayor, Alderman Joseph Barrett[21].

While Andrew Carnegie and his wife were invited to the official opening of The Carnegie Free Library in Cork, they were unable to attend due to his engagements in Scotland.

The building, designed by city surveyor Henry A. Cutler,[22] was, as reported in the *Cork Examiner* at the time;

> 'Elizabethan in style... the beautiful entrance, balcony and tower lend attractive effects... the interior arrangements ...ensure the best possible utilisation of space. The vestibule... opening into a large open hall, 30 feet by 20 feet...the floor of which is laid in the class of mosaic known as terrazzo executed by Italian workmen. Beyond this hall and in line with the entrance is the lending department, furnished in the most up to date manner. Its dimensions are 60 feet by 30 feet... this part... is lighted mainly from the roof. To the left of the library and occupying equal space with it is the newsroom, lined along the sides with neatly designed newspaper stands, the floor space being occupied by tables for the readers. The corresponding area to the left of the lending department is devoted to the reference library...Flanking the entrance loggia on the ground floor are the librarian's offices on the left and the ladies' reading room on the right, and over these respectively are the Librarian's apartments and the juvenile reading room.
> ... The lending and reference libraries are replete with every convenience necessary for their purpose, the reading room is a model of arrangement, the furniture ...has been executed in Cork. The notice case in the vestibule, beautifully worked in mahogany, the counter in the lending library, also mahogany, the bookstalls and bookcases in the lending library, as well as the counter in the juvenile reading room were all executed by Mr. John Callanan, Cork. The artistic iron railing which surmounts

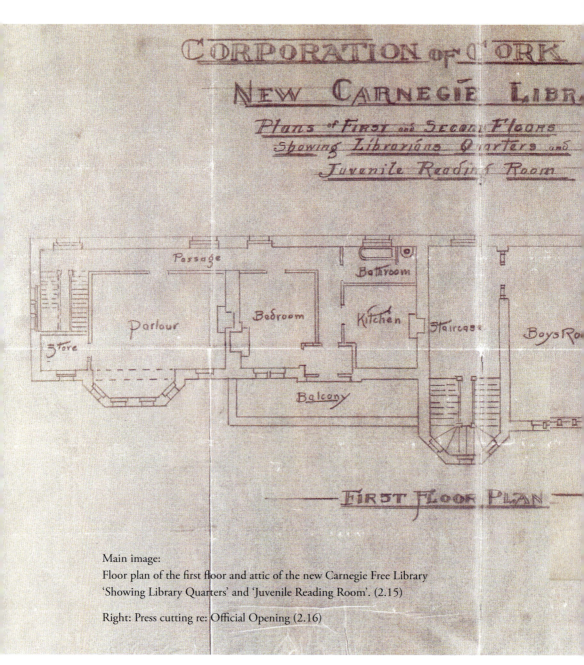

Main image:
Floor plan of the first floor and attic of the new Carnegie Free Library
'Showing Library Quarters' and 'Juvenile Reading Room'. (2.15)

Right: Press cutting re: Official Opening (2.16)

the enclosing walls of the grounds was made by Mr. Watson,
Cork. The carrying out of the contract reflects every credit on
Mr. Patrick Murphy, contractor.'[23]

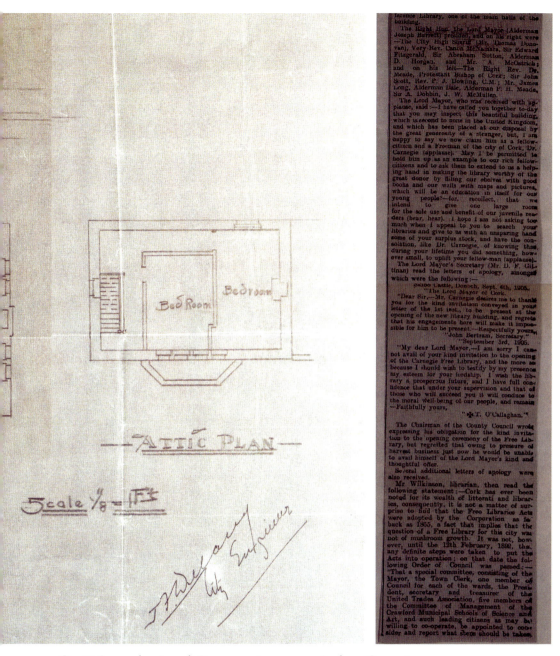

On 13 September Joseph Barrett again wrote to Andrew Carnegie to report on the official opening. He enclosed press cuttings of the proceedings

Lord Mayor's Room.
Municipal Buildings.
Cork.

13th September, 1905

Andrew Carnegie. Esq. LLD.
Skibo Castle.
Dornoch. Sutherland

Dear Sir,

I have much pleasure in acquainting you that the new Library building, which you enabled us to erect for the Citizens of Cork, and which bears your honoured name, was yesterday formally opened to the public by me, in the presence of a large number of leading citizens, and that at a Luncheon given in celebration of the event, your health was most enthusiastically drunk. The feeling of gratitude felt towards you in Cork is indeed warm and universal, and the new Library is admitted on all hands to be at once an ornament to the City and a great boon to our people. I send you herewith copies of the local papers containing a report of the proceedings in connection with the opening ceremonial, and once more tendering you the heartfelt thanks of the people of Cork.

I have the honor to remain, Dear Sir

Yours very faithfully, Joseph Barrett

Lord Mayor

Letter from Joseph Barrett,
Lord Mayor with press cuttings (2.17)

Not all went according to plan however. Perhaps in a foretaste of things to come, as reported in *The Times* the following day, the luncheon given by the Lord Mayor as part of the proceedings was disrupted by the Bishop of Cork, Ross and Cloyne who left early together with many of the guests because the Lord Mayor had omitted a toast to the King's health.[24]

This building lasted for only 15 years, until 1920, when it and its contents were destroyed by fire, along with City Hall and much of the city centre, by British Crown Forces in retaliation for continuing guerrilla activity by the IRA[25].

The burning of the Carnegie Library left the city without a public library service until 1924, when premises were provided on a temporary basis for a library in Tuckey Street. That service was transferred in 1930 to its current location at 57-8 Grand Parade.

Above: Ruined remains of the Carnegie Free Library. December 11th-12th, 1920 (image courtesy of *Irish Examiner*) (2.18)

Cork Free Public Library,

EMMET PLACE,

JAMES WILKINSON,
SECRETARY & LIBRARIAN.

190___

Extract from the Annual Report for 1895

The number of visitors to the Reading Rooms has increased very much during the past year, having risen from a daily average of **478·61** in 1894, to **569·89** in 1895.

This has brought out in a very prominent way the inadequacy of the present News Room, both in point of size and of ventilation, for the reception of such a large number of people. The Committee, acting under the advice of Mr. Arthur Hill, B.E., have done all in their power to improve the ventilation, but with only very moderate success, the fact being that the room is altogether too small for the purpose, and, consequently, when crowded as it is almost every night, and especially on damp evenings, the atmosphere becomes most unpleasant and cannot fail to be injurious to those who remain in it for any time.

The question of providing more ample accommodation, both for the readers and the books, is one that deserves the immediate attention of the Corporation, as already the shelves in the book-room are almost filled, and the counter space, which was none too large at the best of times, has been still further curtailed by the erection of a second Indicator. When it becomes necessary to provide a third Indicator a complete re-organization of the building will become necessary, and, in the meantime, some immediate steps should be taken to relieve the congestion of the News Room and to provide additional shelf-room for the books.

Ditto. 1896.

The News-Room continues to be largely availed of, no fewer than **181,944** visitors having used it during the past 12 months, being a daily average of **590·72**, an increase of **20·85** per cent. on the numbers of 1895.

The Committee think it right to again point out to the Corporation that the accommodation in this department of the Library is quite inadequate for the numbers using the room, and as a consequence the overcrowding, especially in the evenings, is very great. They would draw the earnest attention of the Corporation to this subject, and to the remarks thereon contained in their report of last year.

Ditto. 1897.

The visitors to the Reading Rooms continue to increase, and the inadequacy of the accommodation provided for them is consequently more marked than ever, and the Committee would again draw the attention of the Corporation to the necessity that exists for providing additional room in this section of the Library.

CHAPTER THREE

LIBRARY ANNUAL REPORTS

A MEANS OF COMMUNICATION OR
'CONCOCTIONS OF EXUBERANT IMAGINATION'[26]

During the late nineteenth and early twentieth century, Library annual reports had a variety of uses. Primarily, they were used as a way to communicate with library users. New library stock was listed together with samples of items issued to borrowers in an effort to attract new readers. Lists of donors and donations were seen as acknowledgment of help received during the year[27]. To be listed in the annual report of a public library as a donor, or committee member or in another capacity was to have one's standing in the community acknowledged. Lists of users' occupations were used to highlight the 'open to all' nature of public libraries. While the libraries were at risk of use by less desirable users – 'library loafers' and gamblers using the newspaper reading rooms for 'betting intelligence', annual reports were used to refute any notion that such behaviour was acceptable or prevalent.

It is these narrative sections of the annual reports which are so important to historians now. Libraries of course highlighted their successes but they also mentioned problems and difficulties, whether it was a drop in the fiction borrowing as noted at Bootle[28] or the staffing problems suffered in Leicester in 1888[29]. James Wilkinson used the Annual Report of the Cork Library of 1895 to illustrate the need for a new library in Cork.

The need for additional room was noted in annual reports as was the closure of some reading rooms and the reasons why this was so. Library committees also used their annual reports as opportunities to address readers about problematic aspects of their behaviour, for example, borrowers not washing their hands before picking up a book. The preponderance of fiction borrowed was often a cause for concern as was the loss of books following the introduction of open shelving. The opportunity was also taken to pass information on to readers - for example providing reassuring messages regarding measures taken to disinfect books.

Cork Reports

Each Carnegie Library had at its head a Local Library Committee – this was a necessary condition for the drawing down of a grant from the Irish Advisory Committee of the Carnegie Trust for the development of a library service. As Brendan Grimes (1998) in his history of Carnegie Libraries in Ireland notes, committees varied in size from district to district. For example, in Cahirciveen, Co. Kerry in 1910 the library committee had ninety members while that in Drogheda, Co. Louth around the same time had only eight[30]. These committees tended to be made up of local clergymen, bank managers, national school teachers and local councillors. It was quite common for parish priests to be chairmen of these committees and where this was not the case, Grimes comments, it was usually because Roman Catholicism was not the predominant religion in the district.[31]

At its inception on September 12th, 1905, the Cork Carnegie Free Library Committee consisted of the Lord Mayor as *ex-officio* chairman, fourteen members of the city Corporation as committee members, many of whom were also members of other committees in the city like for example, The Technical Instruction Committee of the Crawford Municipal Technical Institute.[32] Four local businessmen were also members, as were one Catholic and two Protestant clergymen. The President of the city's university was a member and the Librarian acted as the committee's

Carnegie Free Library, Cork.

------ ✿ ------

COMMITTEE.

Chairman :
Right Hon. LORD MAYOR (Ald. H. O'Shea, J.P.), *ex-officio*

Vice-Chairman :
L. A. BEA   , J.P.

Ald. J. Kelleher, J.P.	Coun. M. Egan, J.P.
Coun. J. J. Barry.	Rev. T. Tierney, C.C.
Coun. H. H. Beale.	Rev. F. Sikes, B.D
Coun. D. Buckley, J.P.	J. Moynihan.
Coun. P. H. Curtis, J.P	S. O'Cuill.

P. S. O'H-Eigeartaigh, *Pres.* Gaelic League
S. O'Condealbain, *Hon. Sec.* Gaelic League.

Secretary and Librarian :
Mr. JAMES WILKINSON, F.L.A.

Assistant Librarians :
Misses A. Mathews, M. Doyle, E. Desmond, James Desmond.

Attendant :
J. F. NOLAN.

Library Committee members 1913/14 (3.1)

secretary.[33] The library committees retained a similar make up until 1913/14 when the President and Secretary of the Gaelic League in Cork became members, presumably as a result of the Gaelic Revival taking place in the country at the time.

The Annual Reports for the Carnegie Free Library in Cork ran from April 1st to March 31st each year in line with the tax year at the time. They followed a standard format. Each report was addressed to the Members of the Library Committee and opened with a narrative giving specific information on stock holdings in great detail beginning with the quantity of Library Stock held, any additions either by way of purchase or donation during the year, and noted any books withdrawn from stock. For example, stock for the year ending March 31st, 1906 totalled 10,645 volumes[34] and by March 31st, 1915 which is the final report considered in this work, stock held had increased to 13,177 volumes[35].

Details of books purchased and their total cost followed - £100 10s. 5d was spent in 1906,[36]and 1913 saw the largest amount during the years examined spent on new stock, at £130 5s. 71/2 d.[37]

Additions to stock in the Reference Library are quantified, together with details of the numbers of volumes consulted during the year. The reference library appears to have been a busy place with 13,722 volumes consulted in 1905/06. The librarian comments that in fact the figure would have been greater except that

> '(1) The Reference Library was *closed* to the public on 43 days more than last year; (2) No record is kept of Directories, etc, consulted, as in former years, as they are now placed on open shelves'[38]

The increase in stock in the Lending Library year on year is noted as are the issues for home reading. In 1906, these issues were numbered at 94,554, an increase of 8,200 volumes on the previous year[39]. By 1914, the issue of books for home reading had reduced to 82,539, a net decrease of 3,765 on the previous year– in his narrative the librarian writes

'It is interesting to note that the Fiction issues accounts for all but 349 of the decrease…'

Laborious Work

In his commentary on the Lending Library in 1905/06, the Librarian describes how a class list of the books in the lending section is being prepared, on the Decimal system, work, he comments which 'necessitates the cataloguing of each work *de novo*'[40]. It is not until the report for the year ended 1912/13 that the librarian informs the committee that

'the past year has seen the completion of the scheme of re-classifying the books in the Library on what is known as the 'Decimal Classification'.
He continues
'it is utterly impossible to convey to the lay mind the laborious nature of the work of re- classifying a library…some days, half a dozen books being dealt with…on many none at all'. He continues to explain 'this…will enable a person from this City, who is familiar with the system here, to obtain…a book on a given subject from say the National Library, Dublin, by using the same class number as is used in our local Library… this also applies to Libraries similarly classified throughout the world'[41]

This narrative served a four-fold purpose. It was addressed to the Library Committee and informed them of the good and forward- looking work being done in the Library. It also spoke to the borrowers explaining to them how the new system worked and how their Library now had the same cataloguing system as was used worldwide. In addition, because Annual Reports were sent to other Libraries, it was a way of showing what pioneering work was being done in Cork. Finally, it was an acknowledgement to the staff of their hard work.

Juvenile Library

With the building of the new Carnegie Library in 1905/06, the Juvenile Library (originally called the Boy's Room!) was separated from the Adult Library, a move that proved 'most successful' with an increase of over 2,000 issues on the previous year[42](p.5). The Librarian took the opportunity in the Annual Report of this year to suggest that more children's books be purchased and made available for *'reading in the Library only'*.[43] The same suggestion is made by the Librarian in his report for 1907/08 as he reports that no action was taken on his suggestion two years earlier[44].

The narrative form of the Annual Reports continue each year providing information on the retention of books, of quantities and costs of book binding, of cash receipts for fines, application forms, catalogues and books lost or damaged. In 1907/08 £55 3s 10d was received in fines[45]. This was a substantial sum of money considering that £85 was spent on new books by the library that year[46].

Ladies Reading Room

'Before opening the separate room it was an uncommon thing to see a woman enter the general Reading-room, and whenever one did so, it was with the air of an intruder who felt her position, and who would very soon beat a retreat from what appeared to be an embarrassing situation'[47].

The new Carnegie Library contained for the first time in Cork a 'Ladies Reading Room'. The Librarian comments in the first year that while no specific record has been kept of the use of the room it appears to be well attended and appreciated by the ladies using it[48]. Interestingly, by the following year 1906/07, in his narrative the Librarian expresses 'a considerable amount of annoyance caused by some persons maliciously mutilating some of the journals supplied'[49]

From top: Title Page Annual Report 1913-14 (3.2) and Extract from Annual Report re Ladies Reading Room 1907 (3.3)

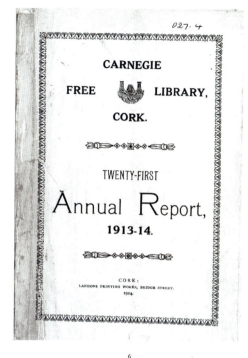

027. 4

CARNEGIE
FREE ⚜ LIBRARY,
CORK.

TWENTY-FIRST

Annual Report,
1913-14.

CORK:
LANDONS PRINTING WORKS, BRIDGE STREET.
1914.

6

to the small attendance during the summer months the Committee decided that the Newsroom shall be closed on Sundays from April to October.

The Newsroom was open on five Bank Holidays, and was visited on those days by 1,963 persons, showing an average of attendance of 392.6.

Ladies' Reading Room.

Although no record is kept of the number of ladies using this room, there can be no question but that it is exceedingly well patronised, but I regret to say there has been a considerable amount of annoyance caused by some persons maliciously mutilating some of the journals supplied. In order to check these propensities the Committee have withdrawn the journals in question from the tables, and have decided to issue them to ladies who are prepared to sign an application form, procurable at the counter. The fact that 1,309 applications have been made for a loan of the journals within the first two months of their withdrawal from the room, emphasises the inconvenience that has been caused to both readers and the staff by the action of a few thoughtless individuals, and at the same time supplies an indication as to the popularity and use made of the journals by interested persons.

Donations.

Donations were received from James Coleman, Esq., Queenstown, to the extent of 14 volumes, as well as the donors named on page 16. The Committee's thanks have already been accorded them.

To Messrs. Egan & Sons (*Cork*) the special thanks of the Committee are due for continuing to keep the clocks in good working order, free of expense.

Staff.

There has been no change in the *personnel* of the Staff, each member of which has performed his or her duty in a satisfactory manner.

The usual statistical tables are appended.

I am, Gentlemen,

Your obedient servant,

JAMES WILKINSON
Secretary and Librarian.

CORK, 4th June, 1907.

As a result of this mutilation the journals were withdrawn from open shelving and application had to be made at the counter for their use. In the first two months of this system, 1,309 applications were made for use of these journals – emphasising, according to the Librarian, 'the inconvenience caused both to readers and the staff by the action of a few thoughtless individuals' and in addition supplying 'an indication as to the popularity and use made of the journals by interested persons'[50]. By the time of the Annual Report of 1909/10, the Librarian noted that 5,777 applications had been made to use the journals and he suggested that in the light of the extra worked caused to staff in the issuing and examination of these item they again be placed on open tables for a trial period[51]. Ladies Reading Rooms will be considered in more detail in Chapter 4.

Miscellanea

Under the heading Miscellanea, the Librarian considered what would probably be termed 'housekeeping matters' today. The Lavatories proved to be a problem in 1905/06 and it was decided to close them against unrestricted use. He suggested to the Library Committee that a charge of One Penny be made for their use, and noted that some libraries receive between £11 and £50 per annum from this source[52]. While that appears to be an eminently sensible suggestion, it wasn't taken up then and there is no mention in further reports of this charge being levied or indeed of the lavatory situation at all.

In his narrative for the first Annual Report of the Carnegie Free Library in Cork, the Librarian noted that the Heating Apparatus which was much lauded in the *Cork Examiner* of September 13th, 1905[53] was not altogether satisfactory and expressed his hope that any defect be remedied[54]. However, it would appear that these repairs were never made or if they were they were not of a satisfactory nature – the heating apparatus in 1907/08 'is still not satisfactory' and 'much too expensive an (sic) one for an institution with such limited means'.[55] The librarian commented that the cost of Heating and Lighting during the past year was £174 10s. 5d., while the amount available for new books was £40[56]. Dissatisfaction with the heating system was again mentioned in 1908/09 – 'No alteration was made to the Heating system, the old costly and unsatisfactory apparatus being used with the same results.[57] Again in 1909/10 he noted that 'The Heating Apparatus still works in an unsatisfactory manner'[58] and in 1910/11,

Carnegie Free Library,
ANGLESEA STREET,

JAMES WILKINSON, F.L.A.
SECRETARY & LIBRARIAN.

Cork, _____ *191___*

*Ann Report
1905-6*

The Carnegie Free Library was opened to the public on the 12th September, 1905, by the Right Hon. the Lord Mayor (Alderman Joseph Barrett), and since that day the work of the institution has gone on in a most progressive manner, the original plans as to administration being carried out without any material alteration.

The Heating Apparatus was not altogether satisfactory during the winter months, but it is hoped that the alterations pending will remedy all defects.

Miscellanea.

*Ann Report
1907-8*

The **Heating Apparatus** is still far from satisfactory, and after two years' experience of the system in use, I have no hesitation in hazarding the opinion that, however efficient it may prove, it is much too expensive an one for an institution with such limited means as obtain here. The cost to the Committee during the past year for Heating and Lighting alone amounted to £174 10s. 5d., while the purchasing power for *new* books amounted to the small sum of £40. I would submit that the former item is too heavy a charge, and that some means be adopted to cut down the expenditure in that respect. With regard to the expenditure on *new* books, it *must* be admitted that the amount named is utterly inadequate to efficiently meet the demands made on the Library, and I would suggest that the Committee endeavour to devise some means of conserving the funds so as to increase the amount available for the purchase of books.

Miscellanea.

*Ann Report
1908-9*

No alteration was made to the Heating system, the old costly and unsatisfactory apparatus being used with the usual results. By using every economy the cost of heating and lighting was reduced by £30, the use of coke in lieu of coal for the furnace proving a considerable factor in that respect.

*Ann Report
1910-11*

8

A satisfactory effort was also made to reduce the introduction of foreign matter in to the building by the heating apparatus, by the fixing of the finest wire gauze across the openings of the inlets ; nevertheless, it is still apparent that the only remedial measure to effectively deal with this nuisance, is the adoption of a filtration scheme ; or, as an alternative, the discarding of the present costly system, and the introduction of some other sysrem of heating.

Miscellanea—

*Ann Report
1911-12*

The Heating Apparatus still works in an unsatisfactory manner, the defects in which are already known to you, consequently, I need not reiterate same.

Extracts from various Annual Reports re: defective heating apparatus (3.4)

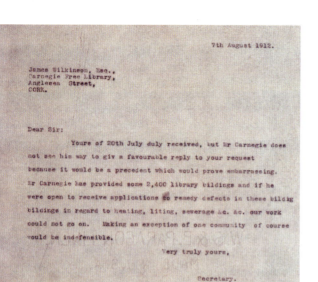

```
                                    7th August 1912.

James Wilkinson, Esq.,
Carnegie Free Library,
Anglesea Street,
CORK.


Dear Sir:

        Yours of 20th July duly received, but Mr Carnegie does

not see his way to give a favourable reply to your request

because it would be a precedent which would prove embarrassing.

Mr Carnegie has provided some 2,400 library buildings and if he

were open to receive applications to remedy defects in these bldig

bildings in regard to heating, liting, sewerage &c. &c. our work

could not go on.  Making an exception of one community of course

would be indefensible.

                                Very truly yours,


                                Secretary.
```

DONORS OF BOOKS.

Donors.				Vols.
American Irish Historical Society	...	...	...	1
Bacon, Mrs. D. G. (*New York, U.S..A.*)		...	...	1
Bannister, W., Esq. (*Watford*)	...	...	...	16
Coleman, J., Esq. (*Queenstown*)	...	...	...	107
Co-operative Wholesale Society, Manchester		...	...	2
Copeman, J. L., Esq. (*Norwich*)	...	...	...	16
Cotterill, J. L., Esq. (*Eastbourne*)		...	...	1
Department of Agriculture and Technical Instruction for Ireland				1
Huggins, Sir W., (*London*)	...	...	...	1
Ingram, J. K. (*Dublin*), Representatives of the late			...	1
Irish Technical Instruction Association, *per* Hon. Secretary			...	1
Invernian Society, Cork, *per* Hon. Secretary		...	...	1
Johnson, Miss E. (*Headley*)	...	...	...	3
Korea ; Resident General for (*London*)		...	...	1
Lawrence, Sir E. D., Bart. (*Ascot*)		...	...	2
Leighton, D., Esq.	...	...	...	1
Lyons, Miss (*Passage West*)		...	...	1
Oakenfull, J. C. (*Plymouth*)		...	...	1
Philadelpha ; Scientific Institution of		...	...	1
Pratt, E. A. Esq, (*Farnborough*)		...	...	1
Scott, G., Esq. (*Dublin*)	...	...	...	1
Smith, H. K., Esq., (*Auckland, N.Z.*)		...	...	1
Smithsonian Institution (*Washington*) *U.S.A.*			...	1
Anonymous	...	...	...	60

LIBRARY REPORTS.—Aberdeen, Ashton-under-Lyne, Aston Manor,
Bangor (*Maine, U.S.A.*), Barrow-in Furness, Battersea, Belfast,
Birmingham, Blackburn, Bolton, Bootle, Bournemouth, Bradford,
Brighton, Bristol, Buxton, Cambridge, Chorley, Cripplegate
Foundation (*London*), Croydon, Derby, Glasgow (*Stirling's*), Glou-
cester. Great Yarmouth, Hammersmith, Hornsey, Hyde, Kettering,
Kensal Rise, Kilburn, Leeds, Leicester, Leyton, Limerick, Middles-
brough, Newcastle-upon-Tyne, New York, Paisley, Perth, Preston,
Rochdale, Saffron Walden, St. Louis (*U.S.A.*), Smethwick, Stock-
port, Tottenham, Wandsworth, Waterford, West Bromwich,
Westminster, Wigan, Windsor, (*Ontario*), Workington, York.

he suggested 'the discarding of the present costly system'.[59] Nothing had been done by the end of the 1911/1912 reporting period – 'The Heating Apparatus still works in an unsatisfactory manner'[60] Mr. Wilkinson wrote to Mr. Bertram, Mr. Carnegie's secretary on 20 July, 1912 asking for a grant towards replacing the unsatisfactory heating system in place. He enclosed cuttings from the Annual Reports to back up his request. Unfortunately for Cork, the reply was in the negative, with Mr.Carnegie unable to help because it would set a precedent of repairing defective buildings, which would prove too costly.

The lists of donors and their donations served as an acknowledgment to those individuals – In 1910/11, the Librarian noted that James Coleman Esq., of Queenstown again headed the list of donors, contributing 107 volumes during the year.[61] The same James Coleman featured among those who generously donated books to the library after the fire in 1920.[62]

While the Annual Reports were, as mentioned before, addressed

From top: Letter from Mr. Carnegie's secretary declining to help with the heating system (3.5) anad List of book donors from Annual Report 1910-1911 (3.6)

to the Library Committee they of were of course also addressing the general public, other librarians and the staff of the Library. The Librarian was careful each year to record the fact that the Staff fulfilled their duties in a satisfactory manner, noting any change in staff, and mentioning with regret the death of staff member Miss Jolley who died in August 1913[63].

The fact that the ongoing difficulties with the 'heating apparatus' and the suggestion of buying new books for the Juvenile Library appears to have been completely ignored by the Library Committee and City Council does seem to fit with some contemporary commentators who claimed that annual reports were a waste of paper[64]. Despite detailed searches, this author was unable to find any mention of the Library Annual Reports in any of the daily or local newspapers in Cork at the time.

Statistical Content

While some of the statistical content of the Annual Reports can appear unnecessary to the modern observer, they are unlikely, as Peatling and Baggs comment, to have been included purposelessly.[65] It is likely that the details of Borrowers' Trades listed, both for males and females, was probably an attempt to show the library committee how widely used the library was and to encourage all members of the public, no matter what, if any, their trade to feel welcome. The Annual reports also contained lists of sample items issued according to the borrowers' trades, this is explained by Ostrum (1990) in Peatling and Baggs (2005) as an attempt to portray public libraries in this period as 'common pool resources and examples of collective action'[66]

The lengthy lists of newspapers, periodicals and journals taken by libraries probably served to publicise the library's stock with the intention of attracting readers. Aside from the book stock both in the Reference and Lending Library, the Newspaper and Ladies' Reading Rooms subscribed to a substantial quantity of both local and national newspapers, journals and periodicals. In 1914 the Newspaper Room took eleven newspapers, fifty four weekly publications, five bi and tri weeklies, forty seven monthlies and eight quarterly publications.[67] In the Ladies' Reading Room, three newspapers and fourteen periodicals were available.[68]

Daily Usage of the Carnegie Free Library

The narrative part of the reports provides a fascinating insight into this aspect of the cultural and educational life of Cork. The opening hours, open seven days a week, including bank holidays 9am – 10pm, evoke a time where the library paid a significant part in the life of the city. Large numbers of people used the Newsroom every day, in 1905/06 the average was 921.75 per day.[69] The decision to close the public lavatories against unrestricted use and the suggestion of a charge of one penny is particularly interesting in light of the difficulties arising in the current Cork City Library because of the lack of public lavatories in the building.

Monthlies—Continued.

*I.C.S. Student.
'Insurance Agent.
*Irish Draper.
Irish Monthly.
Irish Naturalist.
Irish Review
*Journal of the Board of Agriculture (England).
*Journal of the Clerk of Works' Association.
*Labour Gazette.
Lepracaun.
Literary World.
Month.
*Musical Herald.
Musical Times.
Nineteenth Century.
*Patents.
Pearson's Magazine.
*Printer's Register.
Representation
*Return of Deaths of Seamen.
Scribner's Magazine.

Monthlies—Continued.

*Secretary.
*State Correspondent.
Strand Magazine.
*Textile Recorder.
*Universal Patents Bureau.
*Vaccination Inquirer.
*Vegetarian Messenger.
*Vulcan.
*Zoophilist.

QUARTERLY, etc.

American Catholic Quarterly.
Cork Historical & Archæological Journal.
Dublin Review.
Gadelica.
*Journal of the Board of Agriculture and Technical Instruction, Ireland.
Maitre Phonetique.
Progress.
Studies.

Ladies' Reading Room.

*Alliance News.
Cassell's Magazine.
Cork Constitution.
——Examiner.
——Free Press.
Gentlewoman.
Girls' Own Paper.
Illustrated London News.
*Irish Outlook.

Irish Rosary.
*Irish Society.
Lady.
Lady's Pictorial.
Pearson's Magazine.
Queen.
Strand Magazine.
Windsor Magazine.

Juvenile Reading Room.

Boys' Own Paper.
Captain.
Chatterbox.
Children's Friend.

Chums.
Girls' Realm.
Little Folks.
St. Nicholas Magazine.

Railway Time Tables.

*Caledonian (Scotland).
*Cork, Bandon & South Coast.
*Cork, Blackrock & Passage.
*Glasgow and South-Western (Scotland).
*Great Central (England).
*Great Eastern (England).
*Great Northern (England).
*Great Western (England).
*Great Northern of Scotland.
*Great Northern (Ireland).
*Great Southern and Western (Ireland).

*Lancashire and Yorkshire (England).
*London, Brighton and South Coast (England).
*London and North-Western (England).
*Midland (England).
*Midland Gt. Western of Ireland
*Midland & Northern Counties (Ireland).
*North British (Scotland).
*North-Eastern (England).

Visitors to the Newsroom from the 1st April, 1906, to the 31st March, 1907.

MONTH			Visitors	Days Open	Average Daily Attendance
1906					
April	..	..	24,490	24	1020.41
May	..	..	24,417	27	904.33
June	..	..	18,494	26	711.30
July	..	..	18,837	26	724.5
August	..	..	19,870	27	735.92
September	..	..	18,962	25	758.48
October	..	..	23,377	27	865.81
November	..	..	25,775	26	991.34
December	..	..	22,298	25	891.92
1907.					
January	..	..	26,612	27	985.62
February	..	..	25,296	24	1054.
March	..	..	23,893	25	955.72
Total			**272,321**	**309**	**895.79**

SUNDAY & HOLIDAY ATTENDANCES.

SUNDAYS

MONTH			Visitors	Days Open	Average Daily Attendance
1906.					
April	..	..	772	5	154.4
May	..	..	576	4	144.
June	..	..	201	2	100.5
October	..	..	728	4	182.
November	..	..	1,043	4	260.75
December	..	..	1,468	5	293.6
1907.					
January	..	..	1,205	4	301.25
February	..	..	1,205	4	301.25
March	..	..	913	4	228.5
Total	..	..	**8,111**	**36**	**225.30**

HOLIDAYS

DAY		Visitors		
Easter Monday, 1906	..	319		
Whit-Monday, 1906	..	345	Holidays, open 5.	
August Bank Holiday, 1906		308		
St. Stephen's Day, 1906	..	413	Average Daily Attendance	
St. Patrick's Day, 1907	..	578	392.6.	
Total	..	**1,963**		

The lists of newspapers, journals and in particular railway timetables available speak of a city very much part of the British Empire. Available newspapers included the London *Times*, the *Manchester Guardian*, the *Glasgow Herald* and the *Liverpool Post and Mercury*. Railway timetables were available for Great Eastern (England), Great Eastern (Ireland), Great Western (England), Great Southern and Western (Ireland), Lancashire and Yorkshire (England), London and North Western (England) and Midland (England).[70]

Borrowers and their Trades

Lists of borrowers' trades present an indicative list of those who became members of the library. The Classification of the Borrowers Trades for men from

Facing page, left to right:

List of Titles available (3.7)

Details of Visitors to the Carnegie Library (3.8)

Left:

Lists of Borrowers Trades 1913/14 (3.9)

15

Classification of the Borrowers' Trades, etc., from the 1st April, 1913, to the 31st March, 1914.

— ✷ —

MALES.		FEMALES.	
Accountants	31	Artizans	66
Agents	15	Clerks, Book-keepers	54
Architects, Surveyors	2	Domestic Servants	9
Barristers, Solicitors	3	Housekeepers	49
Building Trades	32	Married Women	291
Clergymen	14	Professional	19
Clerks, Book-keepers	189	Pupils, Students	102
Commercial Travellers	5	Shop Assistants	61
Constabulary	7	Shopkeepers, Vintners	24
Electrical Trades	8	Teachers, Governesses	96
Engineering Trades	31	Not described	375
Gentlemen	3	Juveniles	129
Journalists	8		
Leather Trades	7	Total Females	1,275
Manager	10	— o —	
Manufacturers, Merchants	13	**AGES.**	
Printing Trades	8	Juveniles (under 14 years)	3-6
Pupils, Students	146	From 14 to 20 years	588
Railway Employees	9	,, 21 to 30 years	442
Schoolmasters, Teachers	40	,, 31 to 40 years	1-7
Shop Assistants	30	,, 41 to 50 years	62
Shopkeepers, Vintners	25	,, 51 to 60 years	31
Surgeons, Dentists, Chemists	8	Over 60 years	24
Warehousemen	9	Full age	38
Woollen Trades	15	Not stated	45
Miscellaneous Trades	109	Burgesses (age not required)	32
Not described	145	Individual Borrowers	2,443
Juveniles	247	Non-Fiction Tickets issued	193
Total Males	1,1 8	Total Number of Tickets in force on the 31st March, 1913	2,636

April 1st 1909 to March 31st 1910 includes accountants, architects, barristers, solicitors, clergymen, clerks and book-keepers, constabulary, engineers, farmers, gentlemen, journalists, surgeons, dentists, chemists, students, railwaymen, shop keepers, shop assistants, school masters and miscellaneous trades.

The women's classification includes artisans, book-keepers, clerks, domestic servants, housekeepers, married women, professionals, students, shop assistants, vintners, teachers and governesses.[71] While these lists were intended at the time to show that public libraries served the whole city community, today they show historians how popular the Library was with all sections of people. This popularity can also be highlighted by examining the sample lists of books

Carnegie Free Library, Cork

❖

ANNUAL REPORT OF THE LIBRARIAN

For the Year ended the 31st March, 1915,

TO THE

MEMBERS OF THE LIBRARY COMMITTEE.

❖

GENTLEMEN,

I have the honour to submit my **Twenty-Second Annual Report**, which relates to the work of the Library for the year ended the 31st March, 1915.

Library Stock.

There has been a considerable addition to the stock of Books, as **480** new books and 3 books previously withdrawn from the stock have been added; against these additions however, **37** books reported *out of print* were withdrawn, so that the net additions on the year amounted to **446** volumes.

The total number of volumes in the Library at the close of the year is **13,177**.

Books Purchased.

Although only **648** books were purchased as against 1,095 the previous year, the **new** books purchased exceeded by **190** volumes those of the previous year, consequent on a smaller demand for replacements; the number of new books obtained being **423** as against 233 volumes. The replacements totalled **225** as against 862 volumes. The average cost of the books purchased was 3/3½ each, an increase of 10½d. per volume, due to the fact that new books cost more per volume than replacements, many of the latter being procured at low rates in second-hand condition; also, when purchased in a new condition, they are usually obtainable in cheap re-printed editions. The sum expended on the purchase of books was **£106 13s. 2d.**

Bookbinding.

431 books were re-bound during the year, at a cost of £34 9s. 0d.

Books Issued.

There has been a decided fall in the Book issues, there being a decrease of **8,517** as compared with the issues of the preceding year; the total issues amounted to **85,302**, as against 94,017.

It is not difficult to find causes for so large a deficiency, in view of the extraordinary conditions that have prevailed since the outbreak of the war in August last, which seems to have thrown all European countries into a vortex of unrest and from which no community has been exempt; that being so, it is not unreasonable to assume that when such conditions prevail, this and similar institutions must suffer very materially. Further, the *local* activities that have taken place in the volunteering movement, and the increased number of places of amusement, are also factors towards diminishing the output of libraries, from the effects of which this library is no exception. I submit, these elements may reasonably be accepted as being mainly responsible for the reduced circulation, which is the lowest recorded since the year 1897.

It is with pleasure that I am able to report that the combined issues since the Library was opened on the 3rd July, 1893, to the end of the period covered by this report, has now exceeded **Two Millions,** to be exact, **2,046,925**, or, an average issue of over **93,000** per annum. Undoubtedly, the facilities for reading and study which this civic institution provides, have been well availed of, and must have added very considerably to the enjoyment and happiness of the community, more than it is possible to comprehend, for "no one can read a good and interesting book for an hour, without being the better and happier for it."

Reference Library.

The Reference Library issues totalled **9,011** as against 11,478, a decrease of **2,467**. This decrease, seemingly large in bulk, is not very serious when considered in conjunction with the afore stated conditions and from the view-point of the daily average, which works out at **29.44**, as against 37.26; or, a difference of but **7.82** less issues per day.

borrowed during the year contained in each Annual Report together with a note of each borrower's trade. For example in 1909/10 Witts' *'How to look at pictures'* was borrowed by a Clerk, Coal Sorter, Commercial Clerk, Engineer, Hosier's Assistant, Married Woman, Nurse, Plasterer, Retired Warrant Officer, Scholar, School Boy, Ship's Cook, Spinster, Student Teacher and three people whose trades were not stated.[72]

Historical and Cultural Significance of the Annual Reports

Contextually it is significant that these reports are now being examined one hundred years on during a time of intense retrospection in Ireland. The Annual Reports being considered cover a tumultuous period – the Gaelic revival, the Great War and the lead up to the Easter Rising in 1916. Local historians will be interested in the Annual Report of 1914/15 where the Librarian made reference to the 'decided fall in book issues' which was the lowest recorded since 1897 and blamed this on;

> 'the outbreak of war in August last, which seems to have thrown all European countries into a vortex of unrest... and further the *local* activities that have taken place in the volunteering movement'.[73]

The previous year also saw the President and the Honorary Secretary of the Gaelic League joining the Library Committee.[74] This influence of the Gaelic revival on the Library will be of interest to both local and cultural historians. However, except for these examples, the Librarian makes no mention of happenings outside of the confines of the Library.

CHAPTER FOUR

THE LADIES' READING ROOM

A HAVEN FROM
'LIBRARY LOAFERS AND LOUNGERS'[75]

The Ladies' Reading Room in the Carnegie Free Library in Cork was opened to the public on 12th September, 1905. That day's evening paper, the *Evening Echo* reported the provision of '...a ladies' reading room, provided with literature presumably to their tastes'[76]. Review of the library floor plan shows the room to be smaller than the general reading room; however no mention is made of the actual numbers catered for. In his Annual Report of 1905/06 the Librarian states

> 'No record has been kept of the use made of this room, but from general observation it appears to be very well attended, and is much appreciated by the ladies availing themselves of its use'[77]

Mention is made of the room in the narrative section of the Annual Reports up to and including the period 1908/09 but aside from the commentary on the 'persistent mutilation of the illustrated journals'[78] disappointingly little detail is given, except to say that 'the popularity of this room is well maintained'[79]. There is no further narrative on the room, however lists of titles available in it are included in each report up to the final one considered by this work, 1914/15.

Newspapers and Journals

When the Ladies' Reading Room was opened, the stock consisted of two Cork City daily papers, the *Cork Constitution* and the *Cork Examiner* together with thirteen other periodicals. This certainly compares unfavourably with the eleven daily newspapers, local, Irish, British and international together with forty nine weekly and bi-weekly publications and fifty monthly or quarterly periodicals available in the general newspaper reading room at the same time. This time period 1905-1915 coincides with the rapid growth in availability

The Ladies' Reading Room in Glasgow The Ladies' Reading Room in the Cork Carnegie Free Library, for which no images survive, was similar in design and layout.
Image courtesy of Gerald Blaikie.

of journals targeted specifically at a female readership and may be one of the reasons for the expansion of ladies' reading rooms during the period. Ballaster et al, point to the fact that at least fifty new titles, targeted specifically at women, appeared between 1870 and 1900[80].

Chris Baggs notes the average number of titles stocked in ladies' reading rooms to be twenty six, the Ladies' Reading Room in Cork held thirteen. Comparing his list of the Top Twenty Most Popular Titles (see Appendix 1), with the list

of titles available in the Ladies' Reading Room at Cork, it can be seen that the top five periodicals on Baggs' list were held by the Ladies' Room in Cork – they were *Queen, Girl's Own Paper, Gentlewoman, Lady's Pictorial* and *Cassell's Magazine. The Illustrated London News* was also available in the Ladies' Reading Room and also featured on Baggs' 'top twenty' list. The appearance of the first four titles is to be expected as they were popular periodicals of the time aimed largely at a female audience.[81]

Religious and Temperance Magazines

Baggs mentions the appearance of religious and temperance magazines in his 'top twenty' list of titles available in Ladies' Reading Rooms but he believes that they feature there only because in each case they were donated.[82] Examination of the titles available at the Carnegie Library in Cork bears this out, with *Alliance News (*a temperance magazine) and *Irish Outlook* (a Catholic magazine) being donated, however the other Catholic title available, *the Irish Rosary* was paid for by the library.

General Journals

In common with the Ladies' Reading Rooms considered by Baggs, the Carnegie Library in Cork carried a number of 'General Journals' - *Illustrated London News, Cassell's Magazine, Pearson's Magazine* and *the Strand Magazine.* These titles were not specifically aimed at women but at the general reader and contained what can be considered leisure time reading – articles on current affairs as well as a certain amount of fiction[83].

Absent Titles

As mentioned earlier the Gaelic Revival saw an upsurge in interest in Gaelic language papers, for example, *An Lochrann, Scéal Nua, An Claidheamh Soluis* and *Éire Óg*. While these titles were available in the general reading room they were not available to readers in the Ladies' Reading Room. Perhaps more significantly, also missing were any examples of the suffrage press, titles like *Women's Suffrage, Common Cause* and *Suffragette* or any of the titles particular to Ireland at the time, known collectively today as 'the advanced nationalist press.'[84] This absence is particular striking given the number of prominent women writers in Ireland during this period, writers who encouraged the Gaelicisation of life in all its forms – literature, music, sport, political writing, poetry, even women's fashion. Tom Stokes[85] in his article on the feminist, nationalist press in the lead up to the 1916 Easter Rising in Dublin describes Irish women as being *'amongst the most advanced in the world at the time'*.

Maud Gonne, as a journalist and critic of the press, viewed the press as a primary educator in all countries but in particular she noted its influence in Ireland given *'the want of public libraries in most of the country towns and villages of Ireland, and the want of educational advantages, such are found in free countries'*.[86] According to historian Senia Paseta, these women's writing was instrumental in fusing the link between cultural nationalism and advanced nationalism and allowed the creation of a wider space for women's involvement in nationalist politics.[87]

However recent research on the Cork Suffragist Susanne Day and the Munster

Women's Franchise League has unearthed an interesting comment from their first AGM in 1911 -

'Copies of *Votes for Women* have, through our exertions, been placed upon the tables of the Cork Library and of the Carnegie Free Library, two dozen copies of the paper being also disposed of in the city every week, whilst numbers of our members are now regular subscribers.'

The absence of *Votes for Women* from the list of titles held in the Library Annual Reports was probably because the library simply allowed the MWFL to place the papers there - they were not the responsibility of the library. It does show however, the power of advocacy groups even back then. *Irish Examiner*, 09.11.11 p.8

The *Shan Van Vocht* was the first of these publications, founded by Methodist Alice Milligan and Catholic Anna Johnson (Ethna Carberry) in Belfast in 1896 – it hoped to unify nationalists across class, sex and religion[88]. Publishing articles on many aspects of Irish life, cultural, political and nationalist, it provided a valuable platform for women's writings. Steele identifies its primary function as re-awakening interest in Irish patriotism through expressly feminist and nationalist writings.[89] The magazine folded in 1899 and its feminist message was quickly lost[90]. Alice Milligan herself described the demise of the magazine as being due to 'the unfortunate sectional policy which is shattering the forces of Ireland at the moment'[91]. The magazine was re-launched in March 1899 under editorship of Arthur Griffith with the masculine name of *The United Irishman* and lost its expressly feminist outlook. However, as Steele comments it did continue to be one of the most significant nationalist newspapers of the day[92].

The nationalist feminist monthly *Bean na hEireann* was published from 1908-1911 and described itself as 'the first and only nationalist women's paper'[93]. Maud Gonne was the publisher and Helena Moloney the editor. Stokes identifies the tone of *Bean na hEireann* as being undoubtedly militant. It did not support the suffrage movement in Britain, believing instead that women's emancipation would come with Irish national independence[94]. In 1912, Hanna Sheehy Skeffington and Margaret Cousins launched the militant suffrage monthly *The Irish Citizen*. Published until 1920, it campaigned for the vote for women, for equality between men and women, and for national freedom[95]. Like *Bean na hEireann*, *The Irish Citizen* struggled with which cause to put first, feminism or nationalism, and saw as one of its most important tasks the education of women about their warrior past and the need to spur them into militant action in the present.[96]

The advanced feminist press afforded women at the turn of the century a platform for various forms of nationalist and feminist expression. Today it provides researchers with a wealth of information directly from the leading feminist voices of the time allowing their thinking about nationhood, gender and class during this transformational period to be studied. Their complete absence from the Cork Carnegie Free Library conforms with the prevailing attitude of the time which did not consider women to be serious readers and reflects Cork's position as a city still comfortable with its place within the British Empire.

CHAPTER FIVE
'A LADIES' READING ROOM
PROVIDED WITH LITERATURE
PRESUMABLY TO THEIR TASTES [197]

Library designers in the late 19th and early 20th century considered Ladies' Rooms as an absolute necessity for encouraging respectable women into public libraries[98]. James Duff Brown in 1903 suggested that segregated areas might encourage 'women of a fidgety or timid sort'[99] to use libraries. Thomas Greenwood in 1887 in his book *Free Public Libraries* commented that some libraries felt that certain behaviour exhibited by women in public libraries necessitated a separate area, where they could gossip without being disturbed.[100] These rooms were, in the main, small, comfortable, warm spaces, usually with their own toilet facilities. The description of the Ladies' Reading Room in Cork conforms with this, being a safe, comfortable space, away from the general newspaper reading room with its crowds of men, catering to ladies who, it would appear, were largely interested in news, fashion, interiors, Society and the Royal Family. The titles stocked in Cork were almost exactly the same as those stocked in most other Ladies' Reading Rooms in the United Kingdom at the time – titles which were generally held to be suitable reading for 'ladies' both aspiring and actual at the time. Titles concerned with science, music, history, business and so on, while in theory available to women via the general reading room were not available in the Ladies' Room.

All titles held in the Ladies' Reading Room in Cork were printed and published in Britain, despite the existence of a vibrant nationalist/separatist and feminist press in Ireland. In fact, the Gaelic Revival and the general rise in all things nationalist and Irish appears to have had no impact on what the Library Committee considered suitable reading for the ladies of Cork. There was a complete absence of any nationalist titles at all, let alone those of the advanced nationalist press. We have seen however that while Votes for Women was not listed as a title held by the library, the authorities did allow the Munster Women's Franchise League to place copies on the tables. This absence of nationalist titles is surprising considering the political turmoil occurring in Ireland at the time, however there is no mention in any Library Annual Report of any request for or suggestion that this type of reading material be provided.

This was a time of upheaval, of renewed interest in all things Gaelic, through literature, sport and politics. The editorials, articles, poems and plays contained in journals such as *Bean na hEireann, An Shan Van Vocht* and the *Irish Citizen* challenged people's thinking about nationhood, gender and class during this transformative period. The Carnegie Free Library catered at the time to users who were drawn from the ranks of small shop owners, shop assistants, teachers, clerks and so on – precisely the same group of people as John Borgonova points to in his 2013 publication *The Dynamics of War and Revolution: Cork City 1916-1918*, as having formed the small but very active group of Cork separatists.

Were women in Cork not interested in these titles or were the Library Committee and or James Wilkinson not prepared to provide this type of reading material? Unfortunately neither the annual reports nor the correspondence between the Lord Mayors and Andrew Carnegie can provide an answer to this question. No doubt the fact that James Wilkinson, the City Librarian was a Protestant Englishman and the library committee membership was until 1914/15 drawn from upper middle class religious, academic and business men had some bearing on the material provided but more prosaically, the magazines, papers and journals provided in Cork were the same as those provided in public libraries throughout the length and breadth of Great Britain at the time which suggests that Cork Carnegie Free Library was just another British library.

The research undertaken for this book shows that the Ladies' Room probably does reflect the needs, interests and preoccupations of the majority of women library users in Cork at the time. Women who were happy using a new, comfortable library in a relatively prosperous, provincial British city in a period directly before many of them became radicalised by the horrors of World War 1, before they learned of the terrible deaths of young men, many from Cork at places like Gallipoli and the Battle of the Somme and before the complete political and social change in Ireland begun by the Easter Rising and the subsequent War of Independence.

By the time a replacement library was built on Grand Parade in 1930, Cork was a different city. It was part of a free nation – no longer part of the United Kingdom. The provision of a special room for ladies had become an old-fashioned idea and was not considered best practice by library professionals. Thomas Greenwood in early editions of his guide to building and managing public libraries *Free Public Libraries* was inclined to recommend separate rooms for ladies but sometime later changed his mind saying:

'A separate ladies' room means very often a good deal of gossip, and sometimes it is from these rooms that fashion-sheets and plates from the monthlies are most missed. Ladies need not faint at this statement; but it happens to be unfortunately true.'[101]

Indeed we have seen from the Annual Reports that this type of mutilation of ladies journals was a particular problem in Cork at the time.

Epilogue

Cork Carnegie Free Library did not welcome women as full participants. It followed the contemporary norm of believing that women readers and serious readers were two disparate groups. By having a separate space, women were protected and their respectability maintained, however this separation also served to reinforce the socially constructed hierarchy of library users based on their class and gender. The decision not to provide a Ladies' Reading Room in the new Public Library on the Grand Parade following the burning of the Carnegie Free Library in December 1920 reflected the post-World War 1 desire to foster social stability by eradicating all social distinctions. The Ladies' Reading Room had been consigned to history.

Select Bibliography

Manuscript Sources
National Records of Scotland

Carnegie UK Trust Papers

Newspapers
Cork Examiner

Evening Echo (1905). 'Carnegie Free Library Opening Ceremony'. *The Evening Echo* p.2 Cork: Evening Echo

Annual Reports
Carnegie Free Library Cork (1905/06) *Annual Report of the Librarian* (Report No. 13)

Carnegie Free Library Cork (1906/07) *Annual Report of the Librarian* (Report No. 14)

Carnegie Free Library Cork (1907/08) *Annual Report of the Librarian* (Report No. 15)

Carnegie Free Library Cork (1908/09) *Annual Report of the Librarian* (Report No. 16)

Carnegie Free Library Cork (1909/10) *Annual Report of the Librarian* (Report No. 17)

Carnegie Free Library Cork (1910/11) *Annual Report of the Librarian* (Report No. 18)

Carnegie Free Library Cork (1911/12) *Annual Report of the Librarian* (Report No. 19)

Carnegie Free Library Cork (1912/13) *Annual Report of the Librarian* (Report No. 20)

Carnegie Free Library Cork (1913/14) *Annual Report of the Librarian* (Report No. 21)

Carnegie Free Library Cork (1914/15) *Annual Report of the Librarian* (Report No.22)

Secondary Sources
Journal Articles
Baggs, C. (2005) 'In the Separate Reading Room for Ladies Are Provided Those Publications Specially Interesting to Them': Ladies' Reading Rooms and British Public Libraries 1850-1914 *Victorian Periodicals Review*, Vol. 38, No. 3, Fall 2005, pp. 280-306 The Johns Hopkins University Press. DOI: 10.1353/vpr.2005.0028

Peatling, G.K. and Baggs, C. (2004) Early British Public Library Annual Reports: Then and Now – Part 1. *Library History* Vol.20 pp.224-238.

Peatling, G.K. and Baggs, C. (2005) Early British Public Library Annual Reports: Then and Now – *Part II. Library History,* Vol.21 pp.30-44

Murphy, M (1980) The Working Classes of 19th Century Cork. *Journal Of The Cork Historical and Archaeological Society.* LXXXV (241 & 242) 26-51.

Valiulis, M.A. (1995) Power, Gender and Identity in the Irish Free State. *Journal of Women's History,* Vol. 6 no.4/ Vol. 7 No.1 (1995)

Van Slyck, A. (1991) The Utmost Amount of Effectiv [sic] Accommodation: Andrew Carnegie and the Reform of the American Library. *Journal of the Society of Architectural Historians,* Vol.50, No.4 (Dec., 1991) pp.359-383

Van Slyck, A. (1996) The Lady and the Library Loafer: Gender and Public Space in Victorian America. *Winterthur Portfolio,* Vol.31, No.4 Gendered Spaces and Aesthetics (Winter, 1996) pp. 221-224

Wood, B. (1892) 'Three Special Features of Free Library Work – Open Shelves, Woman Readers, and Juvenile Departments'. *Library,* 4

Books

Ballaster, R., Beetham, M., Frazer, E., and Hebron, S. (1991) *Women's Worlds: Ideology, Femininity and the Woman's Magazine.* Houndmills and London: Macmillan Education Ltd.

Bell, J. (2005). *Doing your Research Project* 4th edn.: Open University Press. Maidenhead and New York.

Black, A (1996) *A New History of the English Public Library: Social and Intellectual Contexts, 1950-1914.* London and New York: Leicester University Press.

Black, A. and Hoare, P. (Eds.) (2006). *The Cambridge History of Libraries in Britain and Ireland.* Cambridge: Cambridge University Press.

Borganovo, J. (2013) *The Dynamics of War and Revolution: Cork City, 1916-1918.* Cork: Cork University Press

Collins, M. E. (1993). *Ireland 1868-1966.* Dublin: The Educational Company of Ireland.

Coulter, C. (1993). *The Hidden Tradition: Feminism, Women and Nationalism in Ireland.* Cork: Cork University Press.

Cullen Owens, R. (2005). *A Social History of Women in Ireland 1870-1970.* Dublin: Gill and MacMillan.

Fahy, A.M., (1993), 'Place and Class in Cork', in O'Flanagan, P. & Buttimer C.G. (Eds.), *Cork History and Society,* Dublin: Geography Publications.

Fallon, C (1986). *Soul of Fire A Biography of Mary MacSwiney.* Cork and Dublin: The Mercier Press.

Flint, K. (1993). *The Woman Reader 1837-1914.* Oxford: Clarendon Press.

Grimes, B. (1998). *Irish Carnegie Libraries: a catalogue and architectural history.* Dublin: Irish Academic Press.

Grix, J. (2004). *The Foundations of Research.* Basingstoke: Palgrave Macmillan.

Higgins, R. (2012). *Transforming 1916* Meaning, Memory and the Fiftieth Anniversary of the Easter Rising. Cork: Cork University Press.

Jackson, A. (2003). *Home Rule An Irish History, 1800-2000.* London: Weidenfeld and Nicholson

Kelly, T. (1977). *A History of Public Libraries in Britain 1845-1975.*
London: The Library Association.

Lowney, E. (1990) From Cathleen to Anorexia: The Breakdown of Irelands.
Cork: Attic Press

Luddy, M. (1995). *Hanna Sheehy Skeffington.* Dublin: Dundalgan Press

McCarthy, Thomas (2010). *Rising from the Ashes.* Cork: Cork City Libraries.

Mac Curtain, (2008). *Ariadne's Thread:Writing Women into Irish History.*
Galway: Arlen House.

Moran, C and Quinn, P. (2006) *The Irish Library scene* in *The Cambridge Histories of Libraries in Britain and Ireland Vol. 111 1850-2000* eds. Black, A and Hoare, P.
Cambridge: Cambridge University Press.

Murphy, J. A. (1993) Cork Anatomy and Essence in O'Flanagan, P. & Buttimer C.G.
(Eds.), *Cork: History and Society,* Geography Publications: Dublin pages 793-813.

O'Mahony, C. (1997) *In The Shadows: Life in Cork 1750-1930.* Cork: Tower Books.

Neeson, G (2001) *In My Mind's Eye – The Cork I Knew and Loved.* Dublin: Prestige Books.

Paseta, S. (2013) *Irish Nationalist Women 1900-1918.*
Cambridge: Cambridge University Press.

Pickard, A.J. (2007). *Research Methods in Information.* London: Facet Publishing.

Ronayne, L. and Mullins, J. (2005). *A Grand Parade: memories of Cork City Libraries 1855-2005.* Cork: Cork City Libraries

Steele, K. (2007). *Women, Press and Politics During the Irish Revival.*
New York: Syracuse University Press.

Tiernan, S. (2012) *Eva Gore-Booth: An Image of Such Politics.*
Manchester: Manchester University Press.

Ward, M. (1995) *In Their Own Voice.* Cork: Attic Press

Yeats, W.B. (1921) 'Easter 1916' in *Michael Robartes and the Dancer*
London: Cassell Publishing

Internet Sources

Clonan, T. (2006) *The Forgotten Role of Women Insurgents in the 1916 Rising,* The Irish Times Dublin accessed from www.arrow.dit.ie/aaschmedart/49 on December 12th, 2012
King, D. E. (2009) *Commemoration Address to Dublin City Council* accessed from www.dublinheritage.ie/media/dcpl_125_text.html
Stokes, T. (2012) A Most Seditious Lot: The Feminist Press 1896-1916. http://theirishrepublic.wordpress.com/tag/the-irish-citizen accessed on November 14th, 2012

Appendix 1

Top Twenty Most Popular Titles in British Ladies' Reading Rooms

Queen

Girl's Own Paper

Gentlewoman

Lady's Pictorial

Illustrated London News

Graphic

Quiver

Woman at Home

Myra's Journal of Dress & Fashion

Punch

Cassell's Magazine

Lady

Lady's Realm

Englishwoman's Review

Ladies' League Gazette

Leisure Hour

Our Own Gazette

Taken from Baggs, C. 'In the Separate Reading Room for Ladies Are Provided Those Publications Specially Interesting to Them': Ladies' Reading Rooms and British Public Libraries 1850-1914' Victorian Periodicals Review, Volume 38, Number 3, Fall 2005, pp. 280-306 (Article) Published by The Johns Hopkins University Press DOI: 10.1353/ vpr.2005.0028

Appendix 2

Letter from Florence Bagley Sherman re: changes in the plans. Page 1

Ex90 281/3/84

92

I Landscape Terrace, Sunday's Well,

Cork, Ireland. May 19, 1903.

Hon. Andrew Carnegie,

My dear Sir;

As an American who has lived over six months in Cork, and who has obseved most carefully the workings of the Free Public Library under its very able librarian, I feel justified in calling your attention to what seems to me most serious defects in its control and management by the City Corporation, and to a condition of affairs that I feel sure will seriously affect the future of the library and the benefit the people of Cork ought to receive from your generous gift to them.

It seems to me not too late to prevent these wrongs from becoming settled and fixed if they were known to you, and you felt at liberty to make some suggestion in regard to them.

You know from the report of the library what good work has been done with the very small sum at the command of the committee. They were promised last Fall the full penny rate to which they are entitled and which is absolutely necessary for them to receive in order to be able to move into the new building. When it was asked for this year in March, it was refused, and they were told they were not to receive it until entering the new *library*, which in all probability means two or three years. The committee also asked the privilege of being consulted as to the site of the new building and the plans for it; a request that strikes one as fairly reasonable for the persons responsible to the public and to the corporation to make. The Lord Mayor declined to consult with them, and not only is the library itself seriously crippled by the unreasonable (and I am told by a barrister here) illegal withholding of the penny r

Letter from Florence Bagley Sherman re: changes in the plans. Page 2

2   93

rate, but they have no control of any alterations that may be made
in the plans, the arrangement of the interior of the building, or
any of the things that are properly part of their business.

For instance, the large and well-lighted apartment planned for
the children, has been cut down to a small, dark, unattractive hole
one might almost say. The immense importance of making a library
a place where boys and girls love to go- where they can read and
study in a bright, cheerful room, is being more and more recognized
in our libraries at home, and it certainly seems a great pity that
in this new building, which for many years to come will be the only
place where the children of Cork can learn any thing of the ~~pleasx~~-
pleasures and consolations that books can give them , there is no
proper provision made for them. It is not the fault of the Librarian or the committee; their one plea has been that no essential
of a good library building should be sacrificed to show or ornamen-
tation. But what can they do when they have no voice in the matter-
— when they are not even allowed to come before the Council!

Years ago, when Justin Winsor was in the Boston Public Library,
I remember his good-humoured complaint that the library was "fumbled by committees." Surely a library "fumbled" by politicians with -
-out knowledge or experience is worse.

The site of the new building seems to have been especially chosen
for the purpose of keeping out the people who need a library most,
and who certainly in Cork have shown their appreciation of a
reading ~~rammkyxx~~ room, poor as it is, by crowding it day after day
and by the large use they have made of the reference library.

268,000 reders have used the news room of the present library this
last year, and probably hardly one of these will be able to avail

EKSP 281/3/84   3   94

themselves of the larger privileges of the new library, because of
its position: "It's not much they care for the likes of us with
their fine building; I'm thinkin' they're tryin'to keep us out of
it, putting it out of the world." This I heard one respectable
looking man say to another in the reference room the other day, and
it is quite true that few workingmen or women in Cork can afford to
spend a penny in money, or half an hour of their time in getting to
a building, where they can only stay a few minutes.

I know your gifts are always without limitations and restrictions,
so that the receivers can use them to the best advantage, and that *This*
not only generous but wise giving has been of great service in the
United States, where too often a gift is so tied up with "riders" as
to render it almost no gift at all. But over here, where the
general public have not yet developed any sense of responsibilty in
public affairs , where there are so few public-spirited, independent
thinking citizens, it is a very different matter.

It is not necessary here to discuss the reasons for this condition
of affairs, but the fact remains, that with fewexceptions, there is
no interest taken in any public movements outside the Church and p
politics.   I would not venture to say this if I did not know that
no one, no matter how wise and far-seeing, who is accustomed to our
American ideas and our library methods, can in the least appreciate
the difficulties connected with the management of a public library
in this country. The theological animosities, the class feeling,
the fear of the rate-payer that some one will use a book in the
library who could afford to buy one outside, the subtle and steady
opposition
xxxxxxxxxxxx of the Catholic church to all institutions not con-
trolled by them, make the task a most thankless one, and certainly
an examination of the catalogues of the Cork library, and the
yearly reports since 1893, are sufficient evidence of the quality s

EXCP 28/3/84

4   9S

and quantity of work done.

The only complaint I have ever heard made by the people against the library was that people who could afford to buy a book, or go to a private library sometimes used it, and the librarian was taxed with the fact that a number of carriages had been seen at the door of the building. It was explained that a meeting of a medical association had been held in a lower room, and that the carriages belonged to the doctors and not to the readers. The meaning of "Public Library" in the true sense is quite unknown here and the public must be educated to it.

It is only fair to state that I have had no conversation with any member of the Corporation, or with the Librarian concerning the affairs of the library. My information has been gathered from the daily papers and from outsiders. The Librarian is evidently a man of discretion and judgment, as well as an enthusiast in his profxxx — fession.

Just as I am ending this letter, the morning paper shows by a short editorial note, the truth of what I have just written.

I am about to leave the city, and I think the value and worth of the library here, is so great, so much needed as an institution that in time will break down the class and creed distinctions that now are so marked, that although I appreciate the work that your secretaries must have to do, I would beg of the one who reads this to bring it before you at your earliest convenience, before the opportunity is lost for righting matters.

Believe me, very faithfully yours,

Florence Bagley Sherman.

To Hon. Andrew Carnegie.

Letter from stone cutters association

Andrew Carnegie Esqr. L. L. D.

Dear Sir

On behalf of the Stonecutters of Cork we tender your our hearty thanks for providing our Municipality with the means of supplying a long-felt public want, and adding to the public buildings of our city one the erection of which is doubly welcome to us at present, as its construction affords us a considerable amount of work at a period when our trade locally suffers from a grave dearth of employment.

Under those circumstances we beg to bring the following facts to your notice. Some months ago plans were submitted to, and approved by, you for this institution. These designs were intended to be carried out mainly in cut limestone. When, however, tenders were obtained for the building it was found that they all exceeded by a few hundred pounds the sum allotted by you. It was apparent that either the balance should be provided from some other source, or that the plans and specifications should be so altered as to bring the cost of their execution within the prescribed sum. The latter course was decided on and was effected by taking out a large proportion of the cut stone work and substituting therefor cheaper material of a less durable and less artistic character. This unfortunately included the most prominant feature of the original design—the main entrance.

While not unmindful of the fact that in works of this nature it may be found necessary to make the question of architectural embellish-ment subservient to that of general utility we regret

exC.028.1/3/34

that in the present instance such a course should have been adopted, for the sake of the small sum involved.

When you see that in the immediate vicinity of our New Library are public buildings which in design and workmanship would do credit to any community we earnestly trust that you will cause instructions to be given that the plans which you approved of should be adhered to, so that the structure with which your name shall be always associated, will during its great future of public usefulness remain "a thing of beauty and a joy forever."

Signed on behalf of the Stonecutters of Cork

R. S. M'Namara
General Secretary

EK9028/3/84

APPENDIX 4

Letter re manufacturing etc in Cork. Page 1

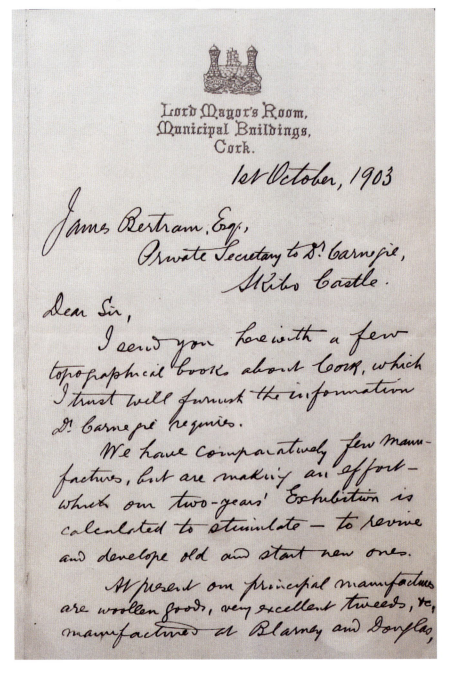

Lord Mayor's Room,
Municipal Buildings,
Cork.

1st October, 1903

James Bertram, Esq.,
        Private Secretary to Dr Carnegie,
                Skibo Castle.

Dear Sir,

I send you herewith a few topographical books about Cork, which I trust will furnish the information Dr Carnegie requires.

We have comparatively few manufactures, but are making an effort — which our two-years' Exhibition is calculated to stimulate — to revive and develope old and start new ones.

At present our principal manufactures are woollen goods, very excellent tweeds, &c, manufactured at Blarney and Douglas,

56

at short distances from the
City. In the City Leather used to
be a staple product, but has
declined, There is also a large
Spinning & Weaving (Linen) Mill,
and large Chemical Works, Artificial
Manure Manufactory, and Several
large Breweries and Distilleries,
which are very prosperous. The
Manufacture of Lace is very successfully carried on in our Art
School and other public schools
(chiefly in connection with the
Convents). Boot & Shoemaking is
another considerable industry.

Butter used to be (and is
still in a lesser degree) the staple
article of trade in Cork, about
Two millions worth at one
time passing annually through our

Letter re manufacturing etc in Cork. Page 3

local Butter Exchange; but the co-operative Dairy System has drawn off to other centres in different parts of the country a large proportion of the trade which used to centralise in the City. Cattle is largely exported. Printing, Bookbinding and other artistic industries are successfully followed. The books I send you herewith are fair specimens of what our principal local publishers can do in this direction, but they can turn out work of still greater beauty & excellence.

An Industrial Development Association has been recently established here, as an outcome of our Exhibition, and is doing good work.

Agriculture is, of course, the main industry of the Country.

I shall be glad to have a line from you by return, as to the amended time table, and Dr Carnegie's final wishes concerning the programme for the 21st.

The Lord Mayor greatly regrets that Mrs Carnegie will not be here to lay the stone.

Yours faithfully
D. J. Gilbman

ENDNOTES

1   Peatling and Baggs, (2000) p.1
2   Black, A. (1996) p.4
3   Peatling, G.K and Baggs, C. 2004 p.224
4   McCarthy, 2010, p.5
5   Fahy 1993, p.790
6   Fahy 1993, p793
7   O'Mahony, 1997, p.316
8   ibid
9   Murphy, 1980, p.30
10 Jackson, 2003, p.143
11 Jackson, 2003, p.144
12 Cullen Owens, 2005 p.22
13 ibid
14 Fallon, 1986, p.22)
15 Paseta, 2013 (p.29)
16 Fitzgerald, E to Carnegie A 26.10.1901
17 Black. A & Hoare, P. 2006 p.253
18 Letter dated 26.10.1901
    Carnegie UK Trust
19 Letter between D. F. Giltinan and
    James Bertram 1.10.1903 Carnegie
    UK Trust
20 Carnegie speech notes, Oct, 21st, 1903
21 Cork Examiner, Sept.13th, 1905 p.3
22 Grimes, B, 1998 p.120
23 Cork Examiner, Sept. 13th, 1905 p.3
24 Grimes, B 1998 p.122
25 McMcCarthy, T. 2010
26 Mason, T (1890) in Peatling and
    Baggs, II 2005, p.40
27 Peatling & Baggs 2004 p.225
28 Borough of Bootle 1891 in Peatling
    and Baggs II 2005, p.35
29 Ibid
30 Grimes, 1998, p.48
31 Ibid p.48
32 Guys Cork Almanac, 1913
33 Evening Echo Sept. 13th, 1905 p.2.
34 CFLC, 1905/06, p.3
35 CFLC, 1914/1915 p.4
36 CFLC, 1906/07, p.2

37 CFLC, 1913/1914, p.2
38 CFLC, 1905/06, p.4
39 CFLC 1913/14, p.5
40 CFLC 1905/06 p.5
41 CFLC 1912/13, pp4-5
42 CFLC 1905/06 p.5
43 CFLC 1905/06 p.5
44 CFLC 1907/08 p.5
45 CFLC 1907/08 p.5
46 CFLC 1907/08 p.5
47 Flint, K 1993 p.108
48 CFLC 1905/06 p.7
49 CFLC 1906/07 p.6
50 CFLC 1906/07 p.6
51 CFLC 1906/07 p.6
52 CFLC 1905/06 p.7
53 Cork Examiner 13.09.05 p.3
54 CFLC 1905/06 0.8
55 CFCL 1906/07 p.6
56 CFCL, 1907/08, p.6
57 CFCL, 1908/09, p.6
58 CFCL, 1910/11, p.6
59 CFCL, 1909/10, p.6
60 CFCL, 1911/12 p.6
61 CFCL, 1910/11 p.7
62 McCarthy, T 2010 p.62
63 CFCL, 1913 p?
64 Maw 1900/01 in Baggs 2005 p.37
65 Peatling and Baggs 2005, p.33
66 Ostrum in Peating & Baggs 2005
67 CFCL, 1914/15 pp.22-23
68 CFCL, 1914/15 pp.22-23
69 CFCL, 1905/06 p.6
70 CFCL, 1905/06 p.22
71 CFCL 1909/10 p.14
72 CFCL 1909/10 p.16
73 CFCL, 1914/15 p.5
74 CFCL, 1913/14 p.2
75 Rae 1899/1900 in Baggs, 2005 p.281
76 Evening Echo, 13.09.05 p.2
77 CFLC, 1905/07 p.7

ENDNOTES

78 CFLC, 1908/09 p.4
79 CFLC, 1908/09 p.6
80 Ballaster et al, 1991 p.75
81 Baggs, 2005 needs finished ref
82 Baggs, 2005 p.287
83 Baggs, 2005 p.289
84 Steele, 2007 p.2
85 Stokes, 2012 p.1
86 Gonne (1900) in Steele, 2007 p.87
87 Paseta, 2013 p.31
88 Steele, 2007 p.65
89 Steele, 2007 p.64
90 Stokes, 2012 p.4 & Steele, p.64
91 Milligan in Steele, 2007 p.64

92 Steele, 2007 p. 67
93 Stokes, 2012 p.4
94 Stokes, 2012 p.5 & Steele, 2007 p.110
95 Stokes, 2012 p.8 & Steele 2007 p.126
96 Steele, 2007 p.126
97 Evening Echo, 12.09.1905
98 Van Slyck (1996) p.221
99 Brown, 1903 in Peatling & Baggs II, p283
100 Greenwood, 1903 in Peatling & Baggs II, p.283
101 Greenwood (1890) in Grimes (1998) p.42

ACKNOWLEDGEMENTS

This book grew out of my MSc Dissertation at Aberystwyth University and would never have appeared without the friendship, help and support of a number of people.
Liam Ronayne, Cork City Librarian was instrumental in bringing this work to print.
Paul O'Regan and Paul Devane of Cork City Library helped with images and other technical issues.
My colleagues at Cork City Library were endlessly supportive and encouraging.
Lucy Smith of the Carnegie UK Trust solved a conundrum and pointed me in the right direction.
Samantha Smart of the National Records of Scotland in Edinburgh provided invaluable help and guidance.
My thanks to Stuart Coughlan of edit+ for his design skills and vision.
I would also like to thank the staff at the National Library of Ireland and at the National Photographic Archive.
This book seeks to add in a small way to the body of feminist scholarship in Ireland and I would like to acknowledge the impact involvement with the Department of Women's Studies in UCC has had on my thinking and writing. In particular, I would like to thank the late Dr. Carmel Quinlan and Dr. Sandra McAvoy for their unfailing support and encouragement.
Finally, as always, I would like to thank my very patient family, my children Daniel, Hannah, James and Katy and my husband Joe.
Kilcrea – May 2015